Earth Alert!

Rivers in Danger

Polly Goodman

Gareth Stevens
Publishing

Please visit our website, www.garethstevens.com.
For a free color catalogue of all our high-quality books,
call toll free 1-800-542-2595 or fax 1-877-542-2596.

Library of Congress Cataloging-in-Publication Data

Goodman, Polly.
Rivers in danger / Polly Goodman.
    p. cm. — (Earth alert!)
Includes index.
ISBN 978-1-4339-6008-6 (library binding)
1. Rivers—Environmental aspects. 2. Stream ecology. 3. Water—Pollution. I.
Title.
GB1201.7.G66 2011
333.91'62—dc22

                          2010049262

This edition first published in 2012 by
Gareth Stevens Publishing
111 East 14th Street, Suite 349
New York, NY 10003

Copyright © 2012 Wayland/Gareth Stevens Publishing

Editorial Director: Kerri O'Donnell
Design Director: Haley Harasymiw

Printed in China

Picture acknowledgements
Shutterstock, main picture; Axiom Photographic Agency (Jim Holmes) 12, (Chris Coe) 25; Clearwater Projects/Allyson Bizer
27; James Davis Travel Photography 24; Ecoscene (Alexandra Jones) 1, (Anthony Cooper) 3, (Frank Blackburn) 9 bottom,
(Anthony Cooper) 9 top, (Alexandra Jones) 16, (Erik Schaffer) 17, (Paul Ferraby) 20, (Jim Winkley) 22, (Nick Hawkes) 23,
(John Wilkinson) 26; Eye Ubiquitous (David Cumming) 11, (L. Fordyce) 15, (David Cumming) 18, 28; Hodder Wayland (David
Cumming) 5, (Julia Waterlow) 6, 13 top, (Julia Waterlow) 14 and 16; Shelagh Whiting 7, 13 bottom; Impact Photos (Robert
Gibbs) 4, (Marco Siqueira) 10, (Michael George) 27; AFP/Getty Images 19.
Artwork by Peter Bull Art Studio.

CPSIA compliance information. Batch #WAS11GS. For further information contact Gareth Stevens, New York, New York at 1-800-542-2595

Contents

What Are Rivers?

Rivers are channels of freshwater that flow downhill toward a lake or ocean. They start as tiny streams, high up in hills or mountains. Rivers are very important because many living things need fresh water to survive.

Waterfalls form where rivers flow over hard rock. ↻

Importance of Rivers

Rivers provide water for washing and cooking, as well as drinking water. Farmers use river water to irrigate their crops, and factories use tons of water every day. People have relied on rivers for transportation for centuries, and they provide a habitat for plants and animals.

Rivers are also a source of power. The force of falling water from rivers and waterfalls can be used to produce electricity.

⌢ **People in India washing in dirty river water.**

The Water Cycle

Only 3 percent of the world's water is fresh, and most of it is frozen in ice caps and glaciers. Only 1 percent is available in rivers, lakes, and underground. This small amount of water is constantly replaced by the water cycle. This is the earth's natural recycling system.

In the water cycle, the earth's water is constantly on the move. Water evaporates from rivers and oceans and turns into water vapor in the air. When it cools, it forms clouds of water droplets, which later fall as rain, snow, or hail. Rivers carry the water down to lakes and oceans. The cycle is happening all the time.

WATER SHORTAGES

Rain does not fall evenly around the world, so many people do not have enough water. More than a billion people in the world do not have clean drinking water.

Parts of a River

The start, or source, of a river is usually a tiny stream. As it flows downhill, the river gets bigger and slower, and is joined by other streams. These streams are called the river's tributaries. The river wears away soil and carries it downstream.

Lower down, the river floods and leaves mud on the flat land beside it. This is called the flood plain.

RIVER FACTS

Longest river: Nile, Africa, 4,145 miles (6,671 km)
Biggest river: Amazon, South America, 4,000 miles (6,437 km). It carries more water than any other river.

The end of a river is called the mouth, where it joins the sea or lake. Some rivers form deltas near their mouth. Deltas are fan-shaped areas, where the river splits up into many channels.

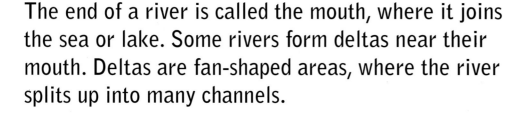

RIVER OF LIFE

The Moruny River is in Kenya. It is home to crocodiles. Leopards, and many other animals and plants, live beside the river. The Pokot people visit the river, with their herds of goats and cattle, for some months every year.

The river provides drinking water for the Pokot people and their herds. The water is also used for cooking. The Pokot people soak a special type of grain in the river until it becomes soft enough to eat.

The Pokot people do not harm the river environment. This is partly because there are not many of them. But they also try not to damage the river because it is so important to them.

⋂ Sandbanks in the Moruny River create shallow pools of water, which are ideal places for herds of animals to drink.

Plants and Animals

Rivers provide a home, or habitat, to many species of plants and animals. They rely on each other for food in a series of food chains.

FOOD CHAINS

Food chains show the paths of food from plants to animals. For example, some plants are eaten by insects. The insects may be eaten by fish, and the fish may be eaten by birds.

Plants

At the bottom of food chains are plants. The most common types of river plant are tiny algae. Algae provide food for fish and insects.

Bigger plants include water lilies and rushes. They provide food and nesting places for animals. Leaves and twigs that fall from plants along the riverbank are food for insects, fish, snails, and worms.

Water lilies on the Amazon River in Brazil. ⮑

Animals

Some animals, such as fish and snails, live in the water. Others, such as birds, take food from the river and live nearby. Frogs spend their time in and out of the water.

⬆ **Otters have webbed feet for swimming.**

The bodies of river animals are specially suited to their habitat. Many have webbed feet for swimming. Swans have waterproof feathers, and crocodiles have streamlined scales.

⬆ **Kingfishers catch and eat fish from rivers.**

RIVER PLANTS AND ANIMALS

PLANTS
Bulrushes
Water lilies
Cattail
Water milfoil
Pondweed

INSECTS
Diving beetle
Water spider
Alderfly
Damselfly

ANIMALS
Stonefly nymph
Freshwater shrimp
Leech
Crayfish
Caddis larva
Flatworm
River limpet

Estuaries

Estuaries are broad mouths of rivers where salt water from the ocean meets the fresh river water. The water in estuaries constantly rises and falls as the tides wash in and out.

A FRAGILE HABITAT

All plants and animals living in rivers are linked in food chains. Damage to one species affects the others in its chain. We must be careful about how we use rivers. Other living things depend on them, too.

Estuaries are home to a huge variety of plants and animals. In the estuary of the Thames River, in the UK, there are shore crabs, water snails, and shrimps. Some plants and animals are specially suited to the estuary habitat. Mangrove trees have long roots to stop the trees being washed away by the tides. Wading birds pluck out insects from the shallow water at low tide.

A wading bird looking for food at low tide in an estuary. ➲

GANGES DELTA

The Ganges River meets the sea in the Bay of Bengal. The river here forms a delta. It splits up into many channels, which flow around 50 islands. This area is called the Sundarbans, after the sundari trees that used to grow there.

There are more mangrove trees in the Sundarbans than anywhere else in the world. There are also many rare animals, including Bengal tigers, crocodiles, spotted deer, and white-bellied sea eagles. In 1966, the area was made a wildlife sanctuary, to protect these species.

The tide has gone out in this estuary, where a crocodile looks for food beside mangrove trees. ➲

Flood

Every river overflows its banks at some time. Some rivers flood more often than others. Floods happen after heavy rainfall, or when snow melts. Rivers near the sea can flood when sea levels rise.

Large floods can ruin crops and houses. But floods can be good for farming. They leave fertile mud on the flood plains around the river.

Flood waters from the Red River in Vietnam flow through the streets of Hanoi. ⟲

People have tried to control floods by building up the banks of the rivers. Special pipes may drain away the flood water. But altering the natural flow of rivers causes problems. If silt settles in the river channel instead of on the flood plains, the river becomes shallow and will flood more easily.

TRUE STORY

THE THAMES BARRIER

The Thames Barrier is a flood barrier in London, UK, near the mouth of the Thames River. It protects the city from floods, caused by high tides from the North Sea.

The barrier has ten separate gates, which can be closed if there is a flood warning. Every part of the machinery has two separate controls, in case one breaks down.

Jan Colby works at the Thames Barrier. He makes sure all the machinery is tested regularly. Jan says: "If there is a flood-tide emergency, the closure team meets in the control room. Within an hour, all the gates are shut."

⋂ The gates of the Thames Barrier, in London, UK.

Jan Colby at work at the Thames Barrier, in London. ⮑

13

Rivers at work

Transportation

Before roads were built, rivers were the easiest way to get around. This is why many settlements started beside rivers. For centuries, they have been used to carry people and goods. Some rivers, like the Danube and Rhine in Europe, flow through many countries from their source to the sea.

Today, in areas where there are few roads, rivers are still the easiest transportation routes. The Amazon River and it tributaries in Brazil carry boats through the dense Amazon rainforest, where it is difficult to build roads. Boats of many sizes travel along the Amazon, from dug-out canoes to huge cargo ships.

These villagers in the Amazon rainforest travel around by boat. ↻

Canals

Canals are artificial rivers. They were built between industrial areas, so that goods and materials could be moved easily from one area to another.

Some canals were created by straightening the sides of natural rivers. Others were dug out and filled with river water.

Activity

LOADING LIMITS

Cargo ships can only carry a limited amount of weight before they sink. On rivers, boats cannot carry as much weight as on the sea because the water is fresh, not salty. Try this activity to see the difference between fresh water and salt water.

1. Collect a plastic lunchbox (without a lid), some stones, and a waterproof marker pen.
2. Float the lunchbox in a sink of freshwater.
3. Fill it with stones until it is just about to sink. Mark the water level on the outside of the box.
4. Now repeat the activity in a sink of salt water. Compare the different water levels. In which type of water can the box carry more weight?

An old mill wheel on the Seine River, in France. ⟳

The tops of dead trees stick out of a reservoir in Australia. ↻

River Power

Rivers have given us power since Roman times. Mill wheels were turned by the rushing water. They drove machines that ground flour and sawed wood.

Today, rivers are used to produce electricity, called hydropower. Dams are built across rivers to make the water drop from a greater height. The water collects behind the dam in a reservoir. When the dam gates are opened, the water falls through machines called turbines. These make the electricity.

However, dams can cause great harm. Their reservoirs flood huge areas that were home to people and animals. When the dam gates open, the sudden flood of water can harm plants and fish downstream.

Drinking Water

Much of our drinking water comes from rivers. It has to be cleaned or boiled to make it safe to drink.

In many places, waste from homes or factories is poured straight into rivers. It can contain poisonous chemicals, oils, and metals, and substances that can cause diseases.

Pollution affects plants, animals, and people who use the river, especially people who cannot clean the water.

Garbage and sewage pollution in a Spanish river. ↻

Tap water is cleaned before it reaches homes. But many people in the world do not have tap water—they drink water straight from rivers.

Industry

Factories can harm the river environment in different ways. Many factories are built beside rivers so that they can transport their goods easily. Some dump their waste straight into the water. Waste from factories farther away from rivers sinks into the ground. Eventually, it seeps into streams and rivers in the ground water.

It is impossible to know if all the poisonous chemicals from industries are removed from our drinking water. Even tiny amounts could be dangerous.

Some industries use large quantities of river water to cool machinery. The water is hot after it has been used. When it is returned to the river, it can kill plants and animals.

⌒ A gasoline factory on the Rhine River, in Switzerland.

DISASTER!

In October 2010, a disaster struck Hungary. Up to 1.3 million cubic yards (1 million cubic meters) of poisonous red sludge from a factory poured out of the reservoir that was supposed to contain it. The flood surged through local villages, destroying homes and killing several people. It poured into rivers that feed the main river, the Danube.

The sludge soon reached the Danube. Hungary's disaster relief services poured huge quantities of materials into the water to counter the effects of the sludge. The Danube is a large river, so the poison was quickly diluted.

But in the Marcal River, the first river the sludge reached, all the fish were killed. No life remains.

Volunteers bring a plastic container of dead fish onto the banks of the Marcal River after the deadly sludge spill. ➲

Farming

Land around rivers is ideal for farming. The soil is fertile, and water from the river can be used to irrigate the land. But if farmers use too much water, wetlands downstream can dry out, destroying valuable habitats.

A river covered in green algae. ↻

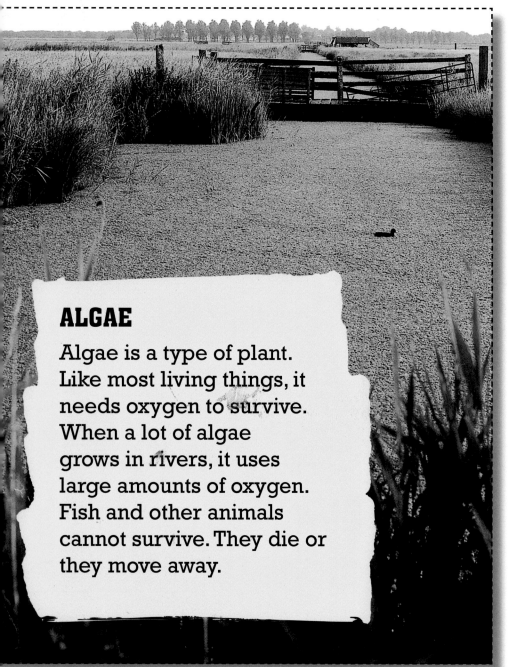

Farmers use chemicals such as fertilizers and pesticides to produce bigger crops. The chemicals sink through the soil and are washed into the rivers. There, they enter the food chains and can damage the river environment.

Animal manure is another type of waste that damages rivers. Manure from cattle, pigs, and chickens seeps into the rivers. The manure makes too much algae grow. This harms the river's habitat.

ALGAE

Algae is a type of plant. Like most living things, it needs oxygen to survive. When a lot of algae grows in rivers, it uses large amounts of oxygen. Fish and other animals cannot survive. They die or they move away.

POLLUTION CHECK

One way to test river pollution is by looking at the creatures that live in it. Try testing a river or stream near you for pollution.

1. Ask an adult to help you collect some water from the river or stream.
2. Use a magnifying glass to study the water. Identify the creatures you see using the diagrams below. If the creatures in your sample are not in the diagrams, use a reference book to identify them.
3. Use the following information to decide whether the river or stream is clean or polluted:

Rat-tailed maggot only: badly polluted water
Bloodworm and rat-tailed maggot: polluted water
Freshwater shrimp or stonefly nymphs: clean water

RIVER SAFETY

* Ask an adult to help you.

* Stay on the riverbank—do not go into the water.

* Avoid steep and slippery riverbanks.

* Wear waterproof boots.

* Do not drink the water.

CLEAN AND POLLUTED WATER.   CLEAN WATER ONLY

Rat-tailed maggot   Bloodworm   Freshwater shrimp   Stonefly nymph

Tourism and Leisure

Rivers are popular places for tourists and other people to spend their free time. They are often quiet, beautiful areas where there is plenty of wildlife. Fishing, bird-watching, and boating are just some of the activities that take place on or around rivers.

However, some sports can be harmful to the river environment. Motorboats can pollute the water with oil and wear away the banks. Lead weights used for fishing can poison fish and birds if they are accidentally swallowed. Even walking and cycling along riverbanks can wear away the banks and frighten away the wildlife.

◖ **Fishing is a popular riverside sport.**

⋔A popular
riverbank.

Development

As an area gets more popular and the number of
visitors increases, parking lots, cafés, and other tourist
attractions are built. Eventually, tourism can drive away
the wildlife and destroy the beauty that attracted
visitors in the first place.

River environments can be protected for tourists if they
are made into national parks or conservation areas.
These have strict rules to limit new buildings.

THE EVERGLADES WETLANDS

The Everglades is a national park in Florida. It includes a long, wide, shallow river. The river is not more than 2 feet (60 cm) deep in most places.

Alligators, crocodiles, snakes, and other wildlife live in the Everglades. Many tourists visit. They travel through the park on airboats.

Even though the Everglades is a national park, the habitat is still threatened by tourists. The airboats can damage the soil and plants. They often injure rare animals called manatees.

↰ An airboat in the Everglades.

Activity

CONSERVATION DEBATE

It is not easy to protect rivers. If an area becomes a national park, it can affect many people's jobs. Find out about different points of view by taking part in this conservation debate with a group of friends.

Imagine the river in this photo is about to be made a conservation area, where no building can take place. Choose one of the roles below. Take turns saying how you will be affected.

A river in the Catskill mountains, in New York. ◑

Roles:
* out-of-work construction worker

* otter living on the riverbank

* local resident

* local business person

Rivers for the Future

Rivers are important, not just to the plants and animals that live in them, but to the millions of people who use them. But as the world produces more waste and uses more water, river habitats are in danger.

If we do not protect our rivers, then our drinking water, homes, and farming could all be affected in the future. There will be fewer beautiful rivers to enjoy, and wildlife will disappear.

You can help protect river environments by making sure your local river is kept free of pollution. Check for garbage, damage to plants, and signs of pollution. You can report any damage you find by writing a letter or sending an email to your local river authority.

These children are collecting small creatures to test for river pollution ➲

THE CLEARWATER PROJECT

The *Clearwater* is a 105-foot (32-meter) sailboat in New York State. It takes people on trips up the Hudson River, to collect and study the wildlife.

Passengers use nets to collect plants and small creatures. After they have studied them, they quickly release them back into the river.

Allyson Bizer works on the boat. She says: "Thousands of people come on the boat each year. It is my job to teach them about the Hudson. The visitors on the *Clearwater* need to learn about how rivers become polluted. They also need to find out how they can help to control problems in the future."

Allyson Bizer on the *Clearwater*. ⟳

A Balance of Needs

Rivers can only be protected by balancing the needs of people and the natural environment. We cannot simply stop developing farming, industry, and tourism. But we can carefully control development so it doesn't destroy the natural environment.

↻ **Tourists relax on a river.**

SNAKES AND LADDERS

Play this game to find out some of the ways people affect rivers. You will need a dice and a counter for each player. Up to four people can play the game. You must throw an exact number to finish.

28

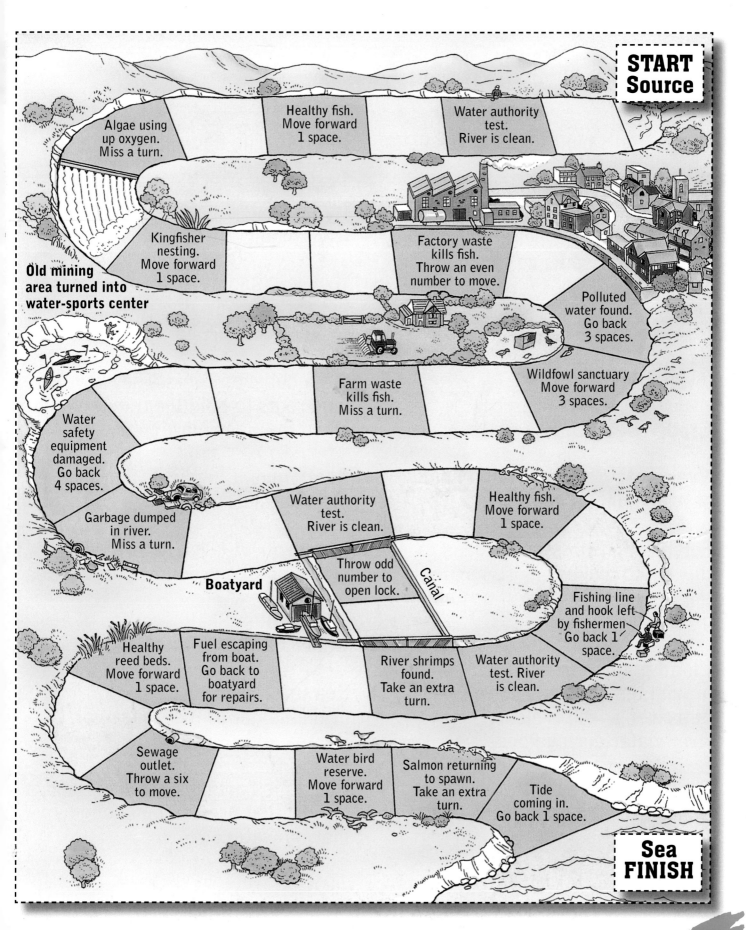

START
Source

Algae using up oxygen. Miss a turn.

Healthy fish. Move forward 1 space.

Water authority test. River is clean.

Kingfisher nesting. Move forward 1 space.

Factory waste kills fish. Throw an even number to move.

Polluted water found. Go back 3 spaces.

Old mining area turned into water-sports center

Farm waste kills fish. Miss a turn.

Wildfowl sanctuary Move forward 3 spaces.

Water safety equipment damaged. Go back 4 spaces.

Garbage dumped in river. Miss a turn.

Water authority test. River is clean.

Healthy fish. Move forward 1 space.

Throw odd number to open lock.

Canal

Boatyard

Fishing line and hook left by fishermen Go back 1 space.

Healthy reed beds. Move forward 1 space.

Fuel escaping from boat. Go back to boatyard for repairs.

River shrimps found. Take an extra turn.

Water authority test. River is clean.

Sewage outlet. Throw a six to move.

Water bird reserve. Move forward 1 space.

Salmon returning to spawn. Take an extra turn.

Tide coming in. Go back 1 space.

Sea FINISH

29

Glossary

Dams Walls built to hold back water.

Deltas Fan-shaped areas near the mouths of rivers where mud settles and the river splits up into many channels.

Downstream The direction a river is flowing, from its source toward the sea.

Erode Wear away.

Evaporate Change from a liquid or a solid into a gas.

Fertilizers Substances added to the soil to make it more fertile. If the soil is fertile, it will be able to grow crops.

Flood plains Areas of flat land around the middle or lower part of a river, which are covered by river water during floods.

Food chains Groups of living things that all rely on each other for food.

Habitat The natural home of plants and animals.

Hydropower Electricity produced from the power of falling water.

Irrigate Supply water to farmland to help crops grow.

Mangrove trees Trees found near the mouths of rivers. They have long roots to hold them in the mud when the tides move in and out.

Pesticides Chemicals that kill insect pests.

Reservoir A large natural or artificial lake used to collect water.

Sandbanks Shallow areas of a river where mounds of sand pile up.

Sea level A measurement of the height of the surface of the sea.

Silt Fine particles of earth, sand, or clay that are carried along by moving water.

Further Information

Topic Web

MUSIC
- *The Moldau* by Smetana
- Handel's *Water Music*

GEOGRAPHY
- Physical features of rivers
- River profile
- Tourism and recreation
- Environmental issues: e.g. erosion, pollution
- Conservation
- Atlas skills: plot the course of a major river.

HISTORY
- Exploration linked to rivers
- Water transportation
- Industrial Revolution and water power

ARTS & CRAFTS
- Using rivers as a stimulus for drawing/painting
- Design a poster for river conservation

DESIGN AND TECHNOLOGY
- Waterwheels and mills
- Design of boats
- Dams and locks

MATH
- Data collection
- Measurement of capacity
- Tally chart of riverboats, barges, etc.

SCIENCE
- States of water: solid, liquid, and gas
- Forces: gravity
- Water cycle
- Floating and sinking
- Biology: flora and fauna, food chains
- Environmental issues: e.g. water contamination

ENGLISH
- Story: *Wind in the Willows*
- Creative writing
- Appropriate poetry
- Library skills

Books

A River Ran Wild: An Environmental History by Lynne Cherry (Sandpiper, 2002)

Landform Top Tens: The World's Most Amazing Rivers by Anita Ganeri (Heinemann-Raintree, 2009)

Protecting Food Chains: River Food Chains by Rachel Lynette (Heinemann Library, 2010)

World in Peril: Rivers under Threat by Paul Mason (Heinemann-Raintree, 2010)

Websites

U.S. Fish and Wildlife Service
http://www.fws.gov/
Search for your local river and find out all about its wildlife, habitats, and refuges.

USGS—Water Science for Schools
http://ga.water.usgs.gov/edu/earthrivers.html
Find out more about rivers, streams, and the water cycle.

National Park Service
http://www.nps.gov/miss/forkids/beajuniorranger.htm
Become a junior ranger on the Mississippi River!

Index